Nauam Oliveira
Fabiano Silva

SOCIAL ENGINEERING IN THE DIGITAL AGE

Nauam Oliveira
Fabiano Silva

SOCIAL ENGINEERING IN THE DIGITAL AGE

MITIGATING RISKS FOR THE ELDERLY

ScienciaScripts

Imprint
Any brand names and product names mentioned in this book are subject to trademark, brand or patent protection and are trademarks or registered trademarks of their respective holders. The use of brand names, product names, common names, trade names, product descriptions etc. even without a particular marking in this work is in no way to be construed to mean that such names may be regarded as unrestricted in respect of trademark and brand protection legislation and could thus be used by anyone.

Cover image: www.ingimage.com

This book is a translation from the original published under ISBN 978-620-6-76096-2.

Publisher:
Sciencia Scripts
is a trademark of
Dodo Books Indian Ocean Ltd. and OmniScriptum S.R.L publishing group

120 High Road, East Finchley, London, N2 9ED, United Kingdom
Str. Armeneasca 28/1, office 1, Chisinau MD-2012, Republic of Moldova, Europe
Printed at: see last page
ISBN: 978-620-7-84810-2

DEDICATION

I want to dedicate this work to my parents who were always there when I needed in help It is what allowed me to pursue my dreams.

THANKS

I thank everyone you my teachers, that during those years transmitted their knowledge, experiences, time It is patience for what we would return successful people and professionals.
I would like to thank my advisor, Professor, Fabiano Pontes Pereira for his support, teaching and interest in the development of this work.
I thank The Banking Examiner for the interest It is availability.

"The only way to do something great It is love O what you he does if You haven't found it yet, keep looking, don't settle."

- Steve Jobs

SUMMARY

O gift work he has as goal to present O context from the social engineering as well as its effects on people with limited knowledge of technology. Social Engineering is the ability to gain access to confidential information and confidential data through persuasion techniques, working from psychological manipulation . Such attacks, whether technical or social, are increasingly present in the modern era in which we live. In today's world, research shows that there has been a large increase in access to networks by the elderly and children, considered per many you preferred targets of the criminals digital, as they are the most susceptible to being affected by cyber threats. Therefore, the objective of this work is to present effective measures so that cases recurrent in blows, thefts It is until cloning in data personal decrease As time progresses and together with these facts, emphasize the paramount importance it has for the general population, helping with how to proceed in situations of cyber attack. For so much, he was used as method for collect in data The search via online questionnaire, data analysis revealed the need to pay more attention to elderly It is to create strategies what promote moments in awareness of access to Internet It is security at network, visa what It is in big importance to maintain that public alert to any attempted coup. The result of this work presents middle of study accomplished one better interaction from the population in general with O world digital through good practices on the Internet.

Key words: Engineering Social, Elderly, Security, Data Personal

SUMMARY

1 INTRODUCTION

According to data from the 149 million Internet users in the country, 142 million connect every, or almost every day, it appears that the majority of users access the network exclusively via cell phone, a reality for more than 92 million individuals. (CETIC.BR, 2023).

The Internet is without a doubt the most advanced technological invention that has brought the most benefits to our society, however we need to be aware of the traps that exist within it and for this there is no other way than to know effective measures in order to face situations of risk to which users are exposed to.

With that The Engineering Social wins big emphasis, one turn what involves a technique used by virtual criminals to trick unsuspecting users into sending confidential data. This way, your computers are infected through malware or links to misleading websites.

The term "phishing" originates from the English word "fishing", due to the similarity between the tactics used by cyber criminals and the practice of fishing . In addition from that, This one crime It is O preferred nowadays in between several others existing due yours easy manipulation psychological, since criminals try to get information confidential, as passwords, numbers in card in credit, information banking or others data personal, Pretending to be one entity reliable.

In 1194 B.C at war of Troy, a horse in wood he was left together to the walls in Troy by the Greeks, supposedly as one gift. You Trojans they took the horse within its walls, believing that the supposed gift was a surrender by the Greeks. In view of this, it is understood that the crimes mentioned above involve some gift or donation that brings harm to the recipient, contrary to what was expected. "We shouldn't ask our customers to balance privacy and security. We need to offer them the best of both. Ultimately, protecting someone else's data is protecting us all." (Tim Cook)

In this work, they are discussed to the questions what involve crimes virtual, as well as measures to alert the general population as well as a specific class, the elderly, about cyber risks. A survey was also carried out in data per quite in quiz with O intention in measure O level knowledge about the security risks that misuse of the Internet can cause.

1.1 PROBLEMS IN SEARCH

With O increase of use from the Internet, you attacks cyber if returned most common, since the increase in users who connect permeates from of the 10 years It is followed of the elderly The leave of the 60 years, according to The search carried out by CGI.br - (Internet Steering Committee in Brazil) , the favorite audience of cybercriminals at which enjoy from the vulnerability of people The things free. Second data from the ICT Households - Technology from the information It is communication (2022), it is stated that 67 million Internet users purchased products and services in 2022. In addition from that, more from the half 51% of the interviewed he did consultations, payments or other financial transactions on the Internet in 2022, an increase of 5 points percentages in relationship to the year previous 46%, in addition in relationship to the multimedia activities, such as watching videos, programs, films or series online it was the most prevalent 80% in 2022 against 73% in 2021. Thus, among the multiple virtual scams carried out against the elderly population on social networks, embezzlement stands out, which can be considered one "trap", one turn what O attacker pervades The to deceive your victim to obtain an illicit advantage from the victim, such as selling someone else's property as one's own or issuing a bad check.In a similar way to fraud, there are malicious calls, as scammers pretend to be family members, companies and accounts in general, requesting data. Therefore, we can formulate the following research question: How can we alert elderly people, who have low technological knowledge, regarding the negative impacts arising from vulnerability in the virtual environment?

1.2 JUSTIFICATION

We live in a post-pandemic world where many traditional services needed to adapt to the online world, such as bank branches, healthcare units, commerce, schools and universities. Although this may be a beneficial situation, risks such as the "Greek gift" mentioned earlier in this work are increasingly present in our society. To this end, misinformation and vulnerability cause the majority of digital security problems, but these risks can be avoided or reduced. Like this, such theme he was selected The end in to analyze O process in transformation of the society we are going through and also emphasize those who are most susceptible to being deceived by so-called attackers. Given the advancement of technology in set with growing number in new types in attacks cyber threats and damage caused to society emphasizes the importance of this study.

1.3 GOALS

Next, they are presented you objectives of this research, divided in general and specific.

1.3.1 goal General

The main objective of this study is to deliver multidisciplinary and preventive approaches aimed at mitigating the risks faced by the elderly in Social Engineering, in which the digital sphere permeates a growing expansion of emerging technologies in which elderly people are vulnerable to cybercriminals, thus strengthening and raising awareness safer environments is of fundamental importance.

1.3.2 GOALS SPECIFIC

- Contextualize people from the third age about The Social engineering ;

- Present the types of technical and social attacks in Social Engineering that attackers use to cause such damage;
- cyber risks ;
- Carry out the study of the data collected to analyze what the current needs are in order to improve the knowledge of the unsuspecting and then build a booklet as an action plan;

1.4 STRUCTURE FROM WORK

In Chapter 1, the introduction is demonstrated together with the research problem analyzed, at which discourse about O as O impact at the increase in users on the internet favors cybercriminals. Furthermore, the justification for this topic in question and the general and specific objectives are presented in order to raise awareness among the general population. In Chapter 2, the theoretical framework of the research is presented. This material is organized into sections that represent each of the main topics covered in this study. Section 2.1 explains to the general population about Social Engineering and demonstrates to the population which they are you main methods in prevention against you attacks. Furthermore, later in the section 2.2 and 2.3 is in fundamental importance to present the main social and technical attacks today. Furthermore, section 2.4 emphasizes cybersecurity concepts through key concepts and current best practices. Next, in section 2.5 the focus goes through a specific group, since the targets favorites of the cybercriminals they are they you Elderly. At the Chapter 3, It is The methodology used in this work is presented, through a survey that aims to assess the population's level of knowledge about Social Engineering at the world digital. At the chapter 4, they are explained you details from the search developed, as well as the results obtained and analyzed. Finally, in Chapter 5, the conclusions obtained are demonstrated together with the closure of the subject.

2 REFERENCE THEORETICAL

This chapter presents the concepts and ideas established for this work, with base in studies previously collected, ensuring The integrity of this study.

2.1 ENGINEERING SOCIAL

Undoubtedly, getting older It is part of development human and it happens naturally, being influenced by factors genetics. Of that form, Because it is a process of loss and greater exposure to diseases, aging he can to bring vulnerabilities, be they social, emotional and physics. (Irigaray et al., 2016). It is of fundamental importance to talk about why the elderly It is more susceptible for if to apply blows, one turn what The lack in knowledge in certain tools what could It is they can avoid scams bankers he does with what you cybercriminals if Enjoy from the fragility of that group, consequently, they are able to steal personal data more easily. In this context, according to the Folha Universal website (2023) The survey conducted by the National Confederation of Managers Shopkeepers (CNDL) It is for the Service in Protection to the Credit (SPC Brazil), in partnership with Offer Wise Pesquisas, reveals a notable increase in internet access in between you Brazilians with more in 60 years, criminals adopted the virtual space due to its greater efficiency: its productivity gains are among the Internet's great success stories (COHEN, 2003).

In agreement with Kevin Mitnick famous hacker already deceased in 16 in July in 2023, in his book "The art of deceiving", says "Social Engineering is the use of manipulation, deception It is influence about one individual belonging The one organization, for what accepts a given request." This request may consist of the disclosure of certain information or the performance of a certain task that benefits the attacker. In this way, a hacker, or simply called a "social engineer", involves a person with some type of authority to request this confidential information, with the ultimate objective of penetrating systems. In this context, in the near past, attacks on computers and other IT devices in network had as goal to achieve O bigger number in systems possible It is cause O maximum in damage, one turn what such attacks no they were, at the However, driven by any specific objective. Furthermore, with the evolution of electronic commerce and the World Wide Web itself, a paradigm shift has been observed, as attacks are becoming more complex and targeted. In short, a "social engineer" basically uses the telephone or the Internet to trick people into giving up confidential information. Furthermore, by using these techniques, "social engineers" take advantage of the human tendency to trust in people, leading to the basic principle used by Social Engineering being that humans are the weakest link in security mechanisms.

In the report "Stopping Insider Attacks" (IBM, 2006), it is possible to identify that in today's world, with the increase in security threats, it is of fundamental importance for organizations, regardless of their sector of activity or the market in which they operate, to commit to increasingly in the security of their systems, investing in the creation of better and more sophisticated defenses, as tools to protect systems (such as firewalls, antivirus, biometric access control devices, etc.) are well identified in today's world. "Traditional" threats aim to target vulnerabilities in security devices. network existing at organization It is what he has access to the outdoor such threats They are mainly based on users' ignorance about security policies in general and the real impact of the damage that their behavior could cause. When we talk in Engineering Social, for what one attack you have success, they are more the skills psychological required per part of "attacker" what win highlighted, than technological ones (Mitnick, 2002). In this context, key skills such as connection and similarity – "Putting yourself in someone else's shoes" are of fundamental importance to attackers, as they allow you to create an environment of empathy that is conducive to the exchange of information. Reciprocity - "I owe you a favor" - The search for future benefits based on the favor provided could prove to be a very useful tool for the "social engineer". In short, the "social engineer" does his "job" well when information is extracted without raising any suspicion.

2.1.1 Law 12,737/2012 (Carolina Dieckmann)

This one chapter verse about The Law Carolina Dieckmann no. 12,737/2012, visa which was created after a global actress had intimate photos leaked, after refusing to pay an amount to criminals, the Law was sanctioned on November 30, 2012, published through of Daily Official from the Unity in 03 in December in 2012 It is came in in force on April 2, 2013, establishing the criminal classification of computer crimes. This way, the history of the creation of this law and some considerations about it will be discussed. Furthermore, the ineffectiveness of the Carolina Dieckmann Law will be discussed, since it is of fundamental importance in the regulation of cybercrimes, but has still left something to be desired. With globalization and the expansion of technologies and the use of virtual media, a new category of crimes called virtual computer crimes or cybercrimes emerged, and the lack of specific legislation regulating such crime left users unprotected. In this way we contextualize O case Carolina Dieckmann cited previously, visa what O access to the actress' privacy occurred due to a common situation that any ordinary person is subject to, when taking their personal computer to carry out a repair, the equipment ended up being hacked through their personal email, in addition, before the emergence of the law , invading a virtual environment and stealing personal data was already a crime, but there were no specific rules on the subject.

Law 12,737/12, popularly known as "Carolina Dieckmann Law", received this name due to the repercussions the Law came to protect the legal good of individual freedom and of the law professional secrecy and personnel, impacting Criminal Law, as there was the addition of two articles which are 154-A and 154-B, entitled "invasion of computer device" and also changed articles 266 and 298 which refer to security in the virtual environment, which provide misuse of information It is materials that match The privacy from the person human at the through the internet, such as photos and videos.

O article 154-A provides about the invasion in device:

Art. 154-A. Hack into someone else's connected computer device or no The network in computers, with O end in obtain, tamper with or destroy data or information without the express or tacit authorization of the device user or in install vulnerabilities for to obtain advantage illicit. Penalty – imprisonment, from 1 (one) to 4 (four) years, and fine. (BRAZIL, 2012).

Furthermore, the art. 154-B disposes about the type in action criminal for such crimes:

Art. 154-B. In the crimes defined in art. 154-A, only proceeds through representation, unless the crime is committed against The administration public direct or indirect action by any of the Powers of the Union, States, Federal District or Municipalities or against public service concession companies. (BRAZIL, 2012).

Therefore, despite such additions to the Penal Code, the Carolina Dieckmann Law need to be more good studied, one turn what for to understand improvement your applicability It is in addition to take in consideration The your importance It is effectiveness in protecting the rights to privacy of the human person, as it aims to repress illicit and criminal conduct carried out in the virtual environment, since there are still crimes that do not have criminal classification, which contributes to impunity, thus use from the Internet It is necessary for all The society, but per other side, Cases of cybercrime are increasingly common, such as embezzlement, theft and data breach, extortion, fraud, pedophilia and for this reason there is a great need to improve standards for combating and preventing cybercrime. In short, a better studied law aims to combat such rampant crimes in the virtual world.

2.1.2 Scratchs to those who have knowledge limited

There is no denying that social networks have revolutionized communications between companies and consumers, however, it is also changing the way people interact with friends and family in general. However, just like any tool that is created for a certain purpose, it can also become a weapon against the user if it is not handled responsibly. Furthermore, it is possible to note that social networks are increasingly being used to obtain user information for various purposes, among which we can mention compromising privacy, establishing profile for malicious purposes, obtain preference information such as restaurants and geographic location. Therefore, the risks caused by social engineering to which users are exposed are enormous, ranging from extortion, loss of confidential or trivial information, to the theft of personal information, as this can often cause great damage. irreparable. a growing trend towards smaller, more specific attacks such as Furthermore, in the "Global Security Index Report" (IBM, 2005), it states SPAM It is SCAM or emails false received and when clicked in misleading way negatively impact those with modest knowledge. In conjunction with these aforementioned risks are the damages caused, since not only those with vast knowledge are subject, but also world-class companies. In this context, according to data, the financial loss caused by social engineering in large companies (public and private) from different segments suffered cyber attacks in 2022 and the perspective is that the invasions no be only in the big ones companies, but also in the in small and medium size. In February 2022, Grupo Americanas' websites were offline for several days and, only in May, the company revealed that it had lost R$923 million in sales in the period, a loss of almost R$1 billion. In short, it can be deduced from the research that the damage caused by social engineering is real and worrying. It should also be noted that the user must always pay attention and be suspicious of any emails or news on social networks that appear suspicious, and always keep their antivirus system updated along with searches for the veracity of such information.

2.1.3 Methods effective of protection

There is no doubt that there are many ways for a social engineer to be successful against a victim, but there may be efficient methods to avoid such success. In addition from that, in each situation there is one manner in if to prevent in attacks of cybercriminals. Thus, through studied bibliographies it can be stated:

- Maintaining silence about confidential situations is essential, as well as being one of the first ways to avoid people with bad intentions.
- Avoiding publishing private or company actions on social networks is also a way of hindering the actions of a social engineer, as such attackers observe their target's every step.
- Avoid using the company t-shirt or badge publicly, unnecessarily, in order to avoid attracting attention and encouraging criminal action.
- In the work environment, avoid writing passwords on paper, or keeping post-its visible with sensitive information, in addition to throwing away papers with determined information, in set check if there is The need to completely shred certain documents and throw them in the trash only when you are sure that nothing is relevant.
- Advanced and intrusion-tested systems are an important item to prevent unwanted intrusions.
- Be aware of avoiding placing identification stickers on cars, notebooks, cell phones, among others, in order to make identification difficult .
- When chatting online, whether alone or in a group, the person must always be suspicious of impersonators, who will try to pretend to be someone they are not.
- Always be suspicious of emails that convey a sense of urgency, or that seek to deceive out of curiosity. Never click on links or open attachments from unknown senders and avoid communicating with them.
- On the cellphone: Be suspicious from unknown or private numbers and never fully believe strange instructions or orders, together always be suspicious if someone asks names, dates or generic questions, be alert.
- False kidnapping: If you are informed that there is a kidnapping, remain calm, and check that the person is really safe for

another means of communication, calling authorities or acquaintances. Also, if someone has actually passed away, or if there has been a major tragedy, check before clicking on the link that says it brings photos and videos, trustworthy news sites, and if the news is true, overcome the temptation to click on the link that may be malware.

• Prizes: If you didn't participate in a draw, or don't remember, be careful, nothing is free.

• Always check whether that person really says who they are, in addition, confirm such attributions or whether they are qualified to carry out such actions.

• Fake profile: Is that person really trustworthy? Be wary of new people who have suspicious photos or who have common friends added to them on social networks, guarantee your privacy above all else, some of your friends can add anyone and the fraudster will use this to create false trust.

• By email: Always watch out for extraordinary offers or impossible promotions, these are usually traps to persuade your emotions.

• On social networks: There are chains of campaigns published daily, bringing great commotion with exciting photos that contain a malicious link, avoid participating or check their authenticity first.

There is no doubt that there are endless applicable tips for preventing social engineering, but all of them would not be useful if the user himself was not aware in what you attacks they exist It is what they exist people but intended in doing what is necessary to fulfill their malicious objectives.

2.2 ATTACKS TECHNICAL IN ENGINEERING SOCIAL

2.2.1 Techniques in attacks

Some forms of technical attacks are widely used by social engineers, who seek to deceive or find a security hole. Therefore, these techniques will be presented below.

2.2.2 Phishing

According to Microsoft, Phishing is a way of obtaining confidential data through fake or cloned websites. Furthermore, the way to apply this method is through an email, in which the attacker describes an emergency situation, generally to reach the victim's emotions and stimulate their curiosity. In this context, when clicking on the link the victim will be directed to a malicious website where some secret information can be collected. Furthermore, another way is to send malware via email and the victim will be infected when they click. To exemplify this fraud, we take, for example, The figure 1 which corresponds to a message from a bank requesting information personal at the which with just one click O attacker has access to the desired information.

Figure 1: Phishing – Theft in information

Dear valued customer of TrustedBank,

We have received notice that you have recently attempted to withdraw the following amount from your current account while in another country: £135.25.

If this information is not correct, someone unknown may have access to your account. As a safety measure, please visit our website via the link below to verify your personal information:

http://www.trustedbank.com/general/cusverifyinfo.asp

Once you have done this, our fraud department will work to resolve this discrepancy. We are happy you have chosen us to do business with.

Thank you.
TrustedBank

Source: The Open university, 2023

2.2.3 Pharming

Pharming involves changing the victim's Domain Name Server (DNS) or the machine's hosts file and when the person enters a specific website they will be automatically directed to a website cloned by the scammer, as the website will be identical to the original and the victim will have modest knowledge will hardly realize that when entering your credentials, the information will be immediately passed on to the hacker, such an attack is very sophisticated. DNS is the server that resolves the name of websites, from a user language in "www.teste.com" to an IP address, for example, 200.255.255.27, which the web server will recognize as a valid request. In this context, when this translator of addresses is adulterated, he does causing the example address above to translate to a malicious destination, such as 200.254.125.27. Furthermore, the victim will be directed to the website without realizing it. cloned, at which until same The URL will be yet original, but only O IP address of the website that will be falsified. Therefore, the second front of the attack will consist of phishing, at the which O fraudster will create one page fake, identical

The original, It is You will then obtain the victim's credentials when they enter and send them to the cloned website.

Figure two: Pharming – Poisoning in DNS

Source: Valimail, 2023

Figure 2 demonstrates the pharming attack, where in the first step the attacker will modify the DNS server, whether local or external (Internet Server Provider – ISP), in the second step the user will request a website and the already modified server will return instead from the original website, the fake website, as shown in the third and final step.

2.2.4   Pop- ups

This technique involves many websites using pop-ups to advertise their products and services, in an advertising style, in a small external window linked to the website. This way, hackers use this method to get an unnoticed click from the victim, and when this happens, it will automatically do so. O download in one malware or it will be targeted

The one page false, whose intention will be to collect confidential information.

2.3 ATTACKS SOCIAL IN ENGINEERING SOCIAL

2.3.1 Differences in between attacks technical It is attacks social

The attacks carried out by social engineers can be divided into direct attacks, in which the social engineer will attack the victim directly via a phone call or in person, requiring the attacker to have a lot of insight and creativity for alternative plans. And indirect attacks, caused by malware, fake websites or emails with scams in their links.You attacks social they are characterized by the attacks direct, at which The The victim has direct contact with the attacker, as they are often disguised and have advance information for a more efficient attack. Technical attacks characterized by indirect types go through virtual routes, as email, SMS, websites cloned, among others techniques what they can deceive the user and allow an attacker to succeed in their attacks. To the too much shapes in attacks will be presented.

2.3.2 Vishing

According to the large digital bank Nubank in its newsletter it defines such a complicated name as vishing being a combination of voice (voice) and phishing (fishing). – fraud virtual), at which such attack pervades for the engineer Social at which will use telephone channels to contact the victim, as the attacker may use in tricks as to alter The identification of the number what it is calling. Therefore, this attack is more related to the target providing credentials or information in accounts banking, among others. At the Brazil that attack It is very common by prisoners, inside prisons, where they randomly call and inform that the victim has won a big prize, and to do so they must credit an amount to a number to confirm the winner.

2.3.3 Smishing

In this attack, the attacker pretends to be an organization or person, often stating that the bank account has been blocked or the card has been cloned, then provide one link for one file or page malicious, to where The scam can be carried out when the victim clicks and is directed to malware or provides information on a false page, including their password. In short, this is a very traditional form of scam, the use of SMS on cell phones.

2.3.4 Dumpster diving

Dumpster diving, or searching through trash, is a method widely used in cases where it is necessary to collect more information from victims. This way, the hacker does not even need to touch the corporate network, or carry out an attack against its powerful firewalls, when the organization does not have an effective method to eliminate you documents O hacker it achieves information valuable just sifting through the trash. A lot of information is found in the trash, such as bank account, information in cards in credits, until same credentials in systems, or data of payslips that are discarded without complete destruction, which can then be recovered by people with bad intentions. In addition, there are also falsehood scams ideological, where to the people discard documents with O your CPF or RG, which criminals use to register in stores, request credit cards, among other ways, and generate great damage to the real owners.

2.3.5 Skimming

Skimming is a technique used to clone RFID (radio frequency) access cards. In skimming, the attacker will use his own system to clone this card with a blank card, just by bringing it close to the equipment. Furthermore, many companies use this method to provide access to restricted areas of the organization for their employees, where the employee simply holds up the card to gain access. This way, with the cloned card, the scammer will gain access in several areas restricted from the company, causing one huge dismantling in your area in security, in addition in power replicate O card several times. (Santos, Pitanga Daniel).

2.3.6 Piggybacking

In the scientific article hacking people through social engineering, it defines PIGGYBACKING as a method used when an attacker decides to follow the victim closely when he or she is entering the building or organization, in this way, most likely, he or she held the door or will leave some gap where the scammer will take advantage. . Nessa technique may to be used one disguise or one stereotype appropriate to the situation that favors the success of the operation. Therefore, this is considered one of the best tactics to achieve restricted access in organizations, as you can obtain a lot of information relevant to your objectives within restricted areas.

2.3.7 Engineering reverse social

Reverse social engineering occurs when the fraudster creates a character fictitious flashy or what appears to have one position in authority It is thus being able to intimidate or deceive the victim (Felipe Nascimento, 2018). In this context, widely used in social networks, such as Facebook, where the social engineer will disguise himself and infiltrate among his target's friends with a fake profile and with many similarities in the interests of the victim and several mutual friends, and the mechanism of social network identifies you as someone interesting to add. This way, the victim adding him will create a much stronger bond, as the interest has gone of the victim.

2.3.8 shoulder surfing

Shoulder surfing is one of the most insightful ways social engineers use to collect information from their targets. Such method presents great effectiveness in collecting in data. One classic use is when the person is typing their password at ATM terminals, where the position of the keys is shown, or they are in a public place, using the company's notebook, which displays stickers with the brand, company logo, It is too much information. Of that form, just with one observation attentive and some photos taken at opportune moments, the social engineer will obtain valuable information.

2.3.9 Hoax

Hoax (rumor) is a tactic widely used virtually, where scammers take advantage of situations that generate great general commotion. Furthermore, it is also classified as SPAM on social networks or telephone messages. The big move of hoax it is in to move emotionally The victim in history sad or serious. This way, when the victim innocently accepts the message and clicks on its link, malware can be installed and all of their contacts will receive the same message. Furthermore, the hoax also consists of messages from companies informing what The victim It is winner in one award or prize draw. You engineers social groups are successful in these attacks, as they know in advance some of the victim's particularities and affinities, using psychological tricks to convince and influence, thus achieving the desired result.

2.4 CYBERSECURITY

Cybersecurity is like a branch of information security that deals with information and information systems that store and process data in electronic form, while information security encompasses the security of all forms of data (such as paper files). (STEINBERG, 2021, np). Together, It is stated that the Internet he was designed so that anyone could easily connect to it and start sending data traffic. This open design has not only spurred a wave of innovation, but also made the Internet a platform for attacks of unprecedented scale and scope. The COVID-19 pandemic has drastically affected the way societies communicate or resolve you problems daily in set you governments must to assess to the policies It is practices in place regarding cybersecurity as the world continues to change, as what as The cybersecurity evolved, also evolved The form as she It is measure. In this way, technology has played a fundamental role in keeping people connected, since for the digital age to realize its potential, a reliable and secure cyberspace must be paramount. Furthermore, as the Internet grew exponentially in Brazil, also grew O number in invasions, fraud and theft of personal data. In this context, to guarantee data security in Brazil, the National Council of Justice enacted Law 12,965 or LGPD - General Data Protection Law, which came into force on 11/08/2020. This law is one of the most modern regulations of its kind in the world and establishes corporate responsibility and sanctions for violations. Below are some of the key cybersecurity concepts that help readers understand them.

2.4.1 Key concepts

Below are some key concepts that will help explain this study: Social Engineering: According to Christopher Hadnagy, in his book Social Engineering: The Science of Human Hacking, social engineering is a set of practices used to persuade individuals to carry out actions that favor the attacker. Because it is conducted on an emotional level, there are no security measures that can stop it. Vulnerabilities: Deduced by weakness in a system, network or application that he can to be exploited per an invader for violate The confidentiality, integrity or system availability. In view of this, it is clear that the exploitation of these vulnerabilities he can violate The security of information, allowed like this O no access authorized The one determined system, network or application, is called such invasion as a cyber attack. Hackers: Pervades anyone who dedicates a lot of time, effort and access authorized in one determined system, network or program in computer. Hackers use their knowledge legally and are typically not motivated by malicious intent. In this context, it can be said that the hacker is the technology professional who seeks to work ethically and always act within the law, seeking to create, modify, develop or adapt new functionalities in software and hardware. Ransomware: Second Ransomware book "Defending Yourself from Digital Extortion" define The one class malware what it suits for extort digitally to the victims, making them pay a specific price. Furthermore, this practice is also known as digital kidnapping, since the target is usually the victims' data, which often contains vital information, whether personal or organizational.

2.4.2 Good practices in cybersecurity

In this topic, this objective is to mention conscious practices, together with small routine attitudes in a very current activity, payment via PIX, in order to preserve our information in the virtual world, some examples will be discussed below.

Payment via PIX, a very current method as it is a more practical way It is quick in effect one purchase, people they are joining The that new shape of payment, including to the people elderly. Although people elderly or what no holds knowledge basics of technology could end being victims in profiteers or criminals.

To protect yourself, you first need to think about whether you really need to share your information with everyone. Furthermore, be careful with people or profiles fake ones that pose as official contacts and offer for help you and ask for your personal details. Remember that social networks are based on connections with people you don't know well and who could be valuable contacts in the future. In this context, some routine practices of fundamental importance stand out:

- Ensure that the information you are seeing on the profile someone are really true.
- Be careful with messages that have links. The links may contain malware and, when clicked, your device may be vulnerable to future intrusions.

- Case there is problems with your Bank, company in card in credit, among others, you should be extremely careful about who you contact. Always look for a reliable communication channel. Make sure the profile of these companies is the official channels.

- Never send personal data by email: Your data is valuable, so if you receive an email requesting your personal information, do not respond. You banks It is others companies Never him asked what send data personal as your CPF, address or login details via email.

- Strengthen the security of your social networks, such as two-factor authentication.
- No click in links or reply emails what request your name in username and password. This information may be used to gain access to your account.

2.5 FROM THE PERSON ELDERLY

Without a doubt, aging is a global phenomenon and is part of the process Natural from the life. A old age from the population human has became one reality in most societies. Thus, over the years researchers have developed ways to prolong life and improve the health of the population. According to IBGE - Brazilian Institute of Geography and Statistics in 2023 it is stated that in absolute numbers, there are more than 33 million people aged 60 or over who reside in Brazil. Furthermore, the constantly changing world requires us to adapt to its evolution, which means that crimes on the Internet have increased in recent years, resulting in new legislation with more severe punishments for any crimes committed in the online environment. Furthermore, the elderly It is one new stage from the life, visa what many elderly need in help, attention and care. In this context, in the Brazil, in 1st October 2003, the The country's legislature approved law No. 10,741 called the Statute of the Elderly.

The law provides:

Art. 2nd - A person elderly enjoy in all you fundamental rights inherent to the human person, without prejudice to the protection integral in what treats it is Law, ensuring if him, per law or per others means, all to the opportunities and facilities, to preserve your physical and mental health It is your improvement moral, intellectual, spiritual and social, in conditions of freedom and dignity.

Furthermore, it is of fundamental importance to highlight that the pandemic also caused an increase in the number of elderly people using the internet. According to recent research by CNDL - National Confederation of Retail Managers and SPC Brasil - Credit Protection Service (2021), it deduces that the number of elderly people who access The Internet grow up in 68% for 97%, in addition of what you main means of interaction for this age group are social networks that allow The distraction, the search for news and communication between family members. In this context, the social network most accessed daily by them is WhatsApp, with 92%, followed by Facebook, with 85% and Youtube, with 77%. Of that form, with O increase of number in connections at Internet, such Fraud crimes have become dizzying on the networks. Elderly people are targeted because they are often unaware of online dangers and tend to have less or no knowledge of digital technology, as several obstacles face an elderly person in an online environment.

Therefore, in set with you data studied It is possible claim that in one World digital each turn more computerized It is connected, It is easy for to the people fall victim to these scams, as internet sites, bank accounts and social media profiles were created for fraudulent purposes, creating a fertile environment for crime. In sum, for what you elderly can enjoy in one old age comfortable, It is I need that they have knowledge in measures effective to the combat against you attackers, thus having peaceful and safe days.

3. METHODOLOGY

The research is about studying the damage and scope of social engineering in Brazilian society, as well as knowledge of cybersecurity and its good practices. Furthermore, it aims to study the relationship between crimes committed, often unnoticed, and also the attitudes of users to the detriment of some risk situations. Data collection was carried out using a questionnaire made available virtually, such questions as age, gender and education open the form in order to identify the responding public, then the form is divided into two parts, general and specific questions. In this context, general questions include frequency of internet use, which applications are most accessed by internet users for communication and whether they pose risks to society. In addition, specific questions include the population's level of knowledge regarding the topic of Social Engineering, which victims are most likely to be deceived, sending false emails and SMS, clicking in a curious or misleading way on random attachments contained on websites. browsed, such damage that these clicks can cause and finally if any victim has already suffered virtual data kidnapping, since the attackers pretend to be such a person. Therefore, the objective of this questionnaire is to assess the knowledge and awareness that the current population has about Social Engineering. It is stated that the questionnaire is the investigation technique composed of a set of questions that are submitted to people with the purpose of obtaining information about knowledge, feelings, values, interests and expectations. In this context, constructing a questionnaire basically consists of translating research objectives into specific questions, since the answers to these questions will provide the data required to describe the characteristics of the researched population or test the hypotheses that were constructed during planning. from the search. Furthermore, he was used researches quantitative It is qualitative. Quantitative research focuses on objectivity in conjunction with mathematical language to describe the causes on one phenomenon and the relationships between variables. Furthermore, qualitative research aims to provide greater familiarity with O problem, with views The make it more explicit or The ramp up hypotheses, one as it involves a research approach that studies subjective aspects of social phenomena and human behavior. In short, the objects of qualitative research are phenomena that occur in a specific time, place and culture. The following graphs show the results of the research, through the questionnaire applied via Google Forms, such graphs permeate TCC'S in previous years sincere thanks to Santos, Pitanga Daniel.

Furthermore, before presenting the graphs, it is important to highlight some information collected regarding to the participants of quiz. Participated of study 84 people It is such information permeates the age of each participant, in addition to analyzing the gender of the participants in this study, the type of sex they fall into was asked in the questionnaire, 64.6% said they were female and 35.4% said they were male. masculine. Furthermore, for to analyze the degree of instruction of participants the level of education was asked, where 79.3% said they had higher education and 19.5% were in high school. In short, given this analysis it is possible to worry about identifying the factors that determine or that contribute to the occurrence of the phenomena.

4. DATA COLLECTION AND ANALYSIS OF RESULTS OBTAINED

Graphic 1.1 – Use from the Internet

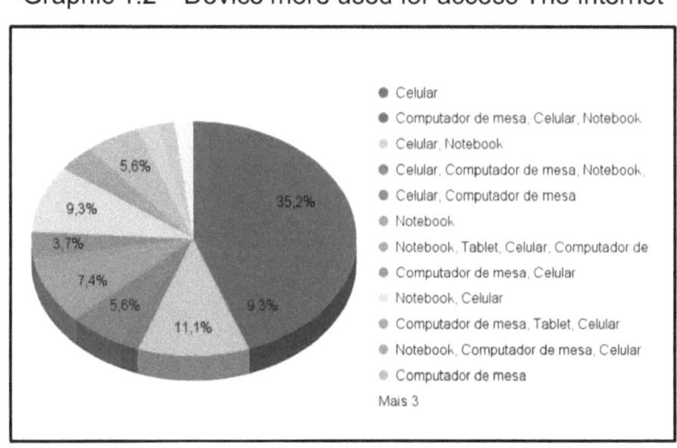

Source: Authorship Own (2023).

Regarding the use of the Internet by respondents, it is possible to see that 98.1% use the Internet several times a day. In this context, it is of fundamental importance to raise awareness of good practices in order to avoid undesirable damage.

Graphic 1.2 – Device more used for access The Internet

Source: Authorship Own (2023).

Regarding the type of device used to browse the Internet, it is possible to notice that there is a dominance of the cell phone device to access the network, of the 98.1% who said they access the Internet several times a day, 35% of the respondents said they use their smartphone as a the main access tool.

Graphic 1.3 – Applications more used for communication

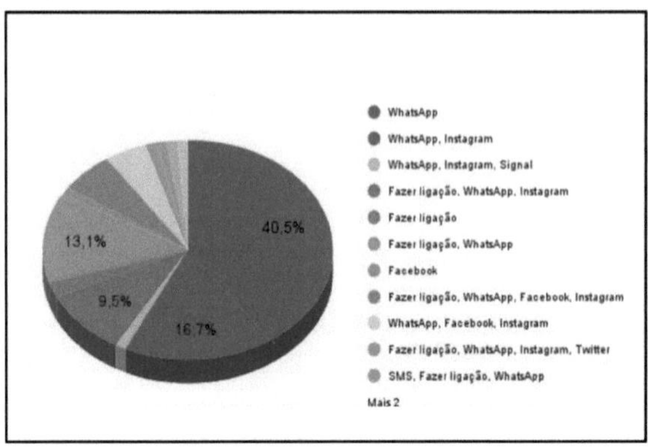

Source: Authorship Own (2023).

Nowadays, very is due to the usability of applications that allow any person can interact easily. Of that form we can to perceive In the graph above, 53.6% of respondents decide to communicate via the WhatsApp application in conjunction with making calls. Furthermore, the social network Instagram it is used by 16.7% of respondents. Given the issues mentioned above, it is important to analyze some situations arising from these two tools mentioned, such as malicious messages or calls.

Graphic 1.4 – Ask help The somebody when he has difficulty in to understand messages or incoming calls

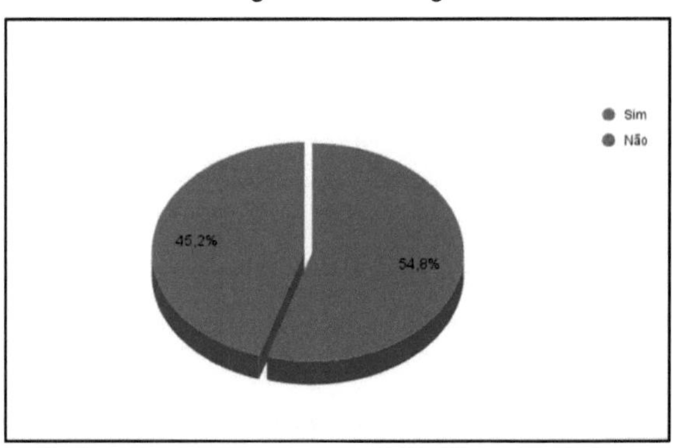

Source: Authorship Own (2023).

In relationship to the respondents ask help The somebody when he has difficulty understanding messages or calls received, 51.9% said yes, they ask for help The somebody what be likely to guide them at outlet in decision. 48.1% of the Respondents state that they do not ask anyone for help, that is, a large proportion are susceptible to risks and deserve greater attention.

Graphic 1.5 – Case somebody call or send one message informing what It is one employee from something Bank It is part your name complete, number of your CPF, number of telephone, number account and password, would you pass it?

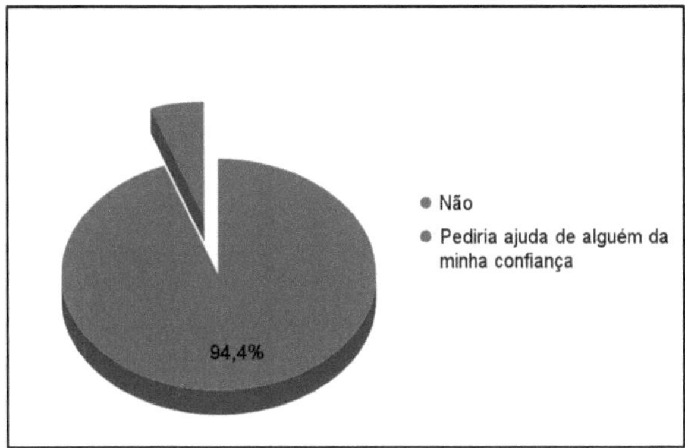

Source: Authorship Own (2023).

Respondents stated that they did not provide any information. The survey showed that 94.4% would not pass on their data, while 6.6% of respondents stated that they asked someone they trusted for help to make a decision.

Graphic 1.6 – O access The Internet offers some danger?

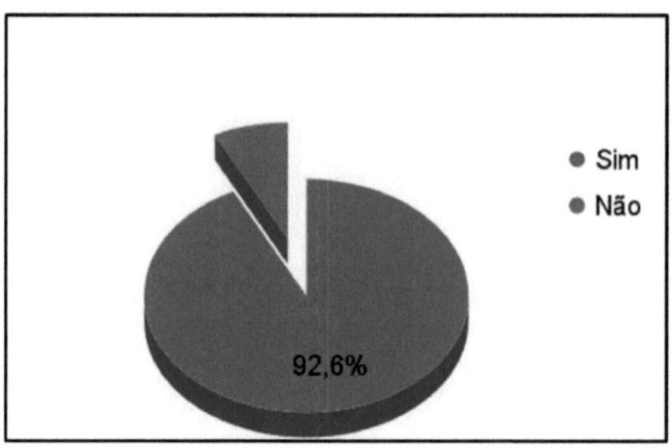

Source: Authorship Own (2023).

As for access to the Internet posing any danger, 92.6% of respondents believe that yes, the internet is a dangerous environment, as anyone who does not understand what circulates in this complex space of the virtual world ends up vulnerable to attackers.

Graph 1.7 – Do surveys like this help raise awareness of good practices in Internet security, in order to identify possible scams?

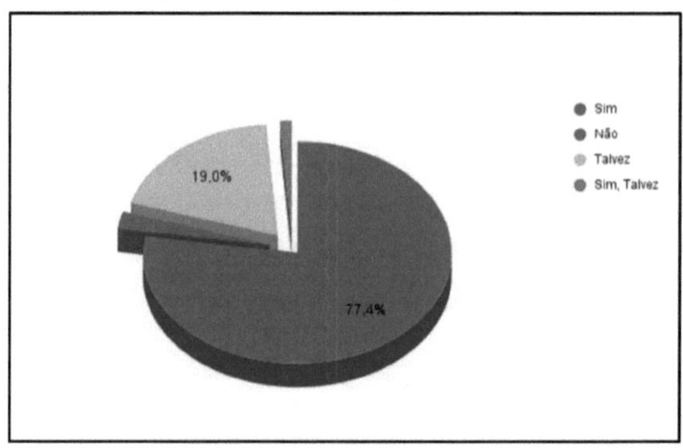

Source: Authorship Own (2023).

How much The realization in researches as such, 77.4% of the respondents believes so, the creation of research helps to raise awareness of good practices, in On the other hand, 19.0% say that perhaps such research could help. Therefore, action plans are necessary in conjunction with research in order to identify scams.

Graphic 1.8 – In your opinion which victims they are more propitious The to be deceived?

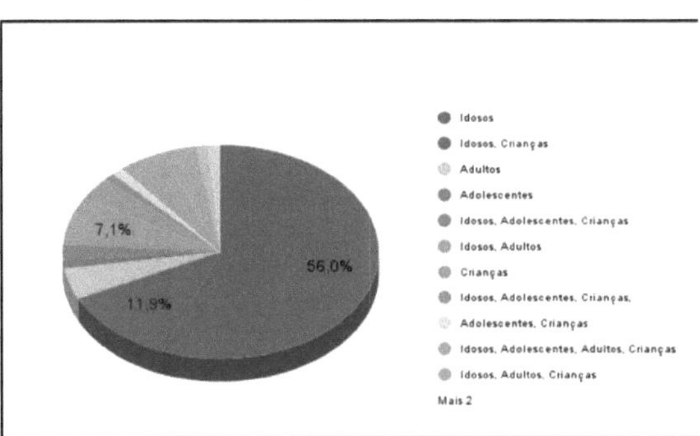

Source: Authorship Own (2023).

It is concluded that the elderly, with 56.0%, are the most likely targets to be deceived, since modest knowledge of technology is the main reason for such a percentage. Next to children with 7.1%. In short, action plans are necessary, together with research, to prevent attackers from taking advantage of such fragility.

Graphic 1.9 – Which your level in knowledge about O theme "Engineering Social"?

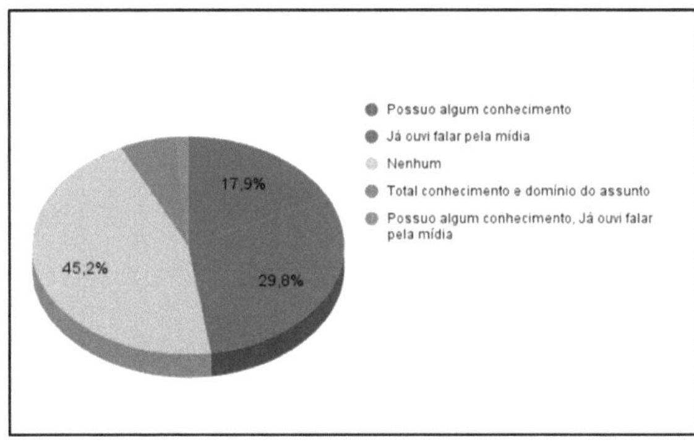

Source: Authorship Own (2023).

It can be deduced that the percentage of people who have knowledge through the media or no knowledge on this topic is 76.8%. Furthermore, 17.9% have basic knowledge and only 4% have complete mastery of the subject. Therefore, the dissemination of research is of fundamental importance in order to avoid the greatest number of scams.

Graphic 2.0 – You already it received one email false?

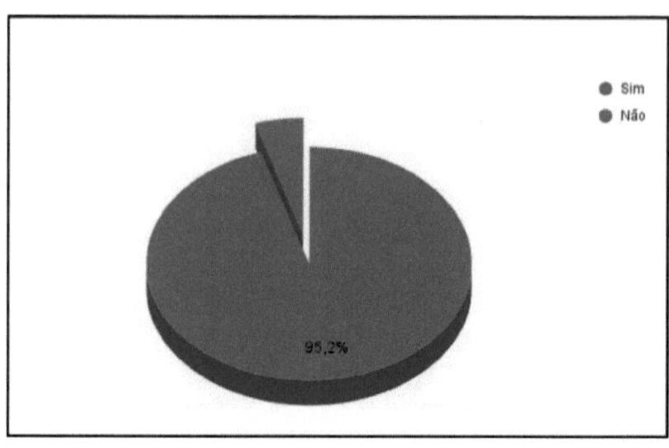

Source: Authorship Own (2023).

Concludes what 95.2% of the respondents already received one email false any time. In this context, it is inferred that this practice is widely used.

Graphic 2.1 – Case you have received one email false: It arrived The to click at the link or attachment contained in it, out of curiosity?

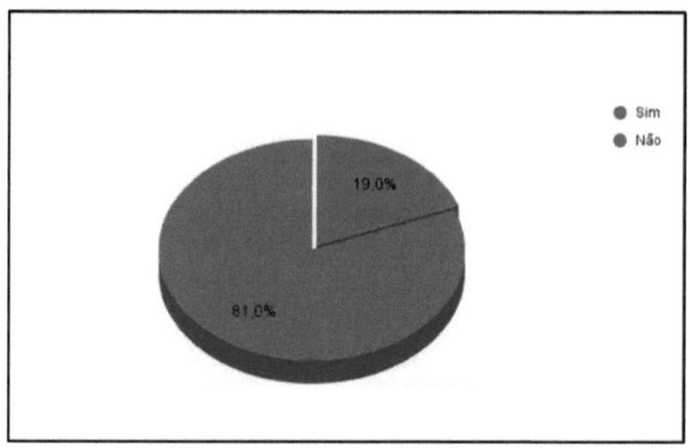

Source: Authorship Own (2023).

With 81.0%, despite receiving fake emails, they do not click on the link or attachment contained, so it can be concluded that the majority of the population is aware of the risks when receiving a suspicious email. However, 19.0% click out of curiosity, a good number for attackers.

Graphic 2.2 – Already clicked in one link in form misleading while sailed in some site?

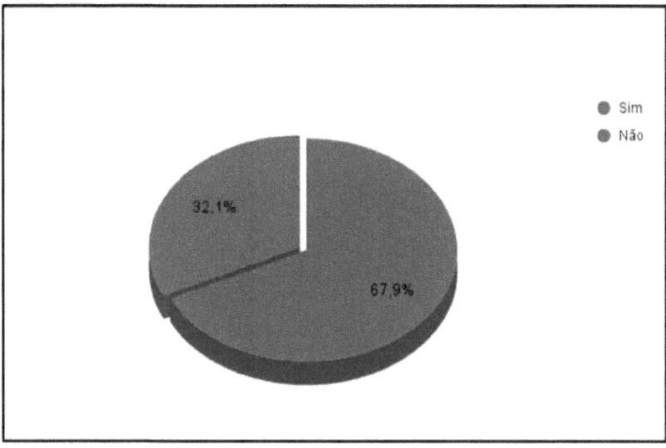

Source: Authorship Own (2023).

The majority, at 67.9%, of users surveyed responded that they accidentally clicked on unknown links while browsing websites. In another side, one good number with 32.1% of the users sail in form careful, in order not to click on individual links.

Graphic 2.3 – Some turn you already it received a phone call with one try in fraud about a kidnapping, prize or bank?

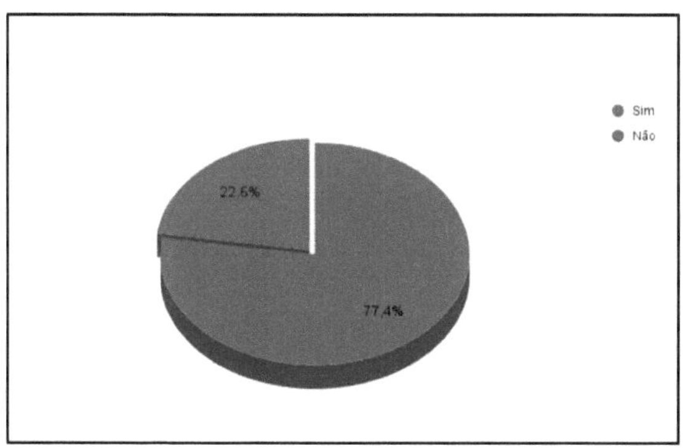

Source: Authorship Own (2023).

The majority of people have received a call from a fraudster, with 77.4% and 22.6% never receiving such a call.

Graphic 2.4 – Case you have received The connection in one fraudster, it arrived The accomplish O what he was order?

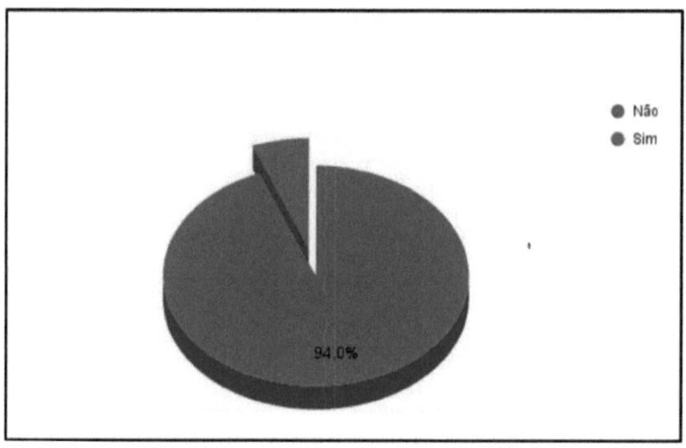

Source: Authorship Own (2023).

As expected, due to the results previous It is for the bigger level in awareness 94.0% did not carry out the request, frustrating the scammer's objectives.

Graphic 2.5 – Already it received SMS false in one prize draw what you have won one award?

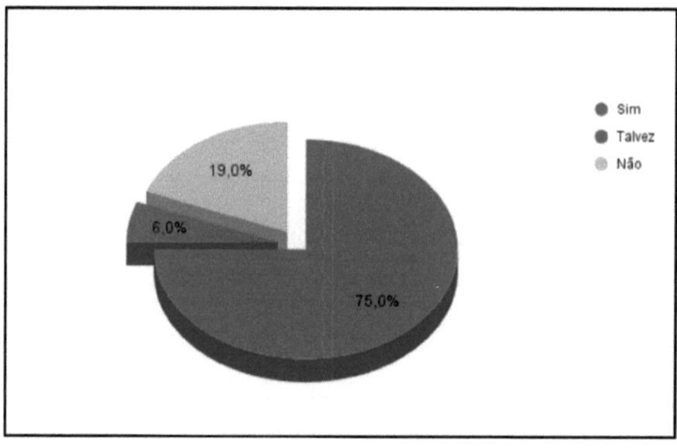

Source: Authorship Own (2023).

In short, 75.0% of participants responded that they had already received a fake SMS, It is 19.0% responded Never to have received. Of that form It is possible to perceive which is a widely used and far-reaching fraud. However, out of curiosity, 6.0% do not know the veracity of an SMS.

Graphic 2.6 – In some time already he had your data cloned, it is card in credit or personal data ?

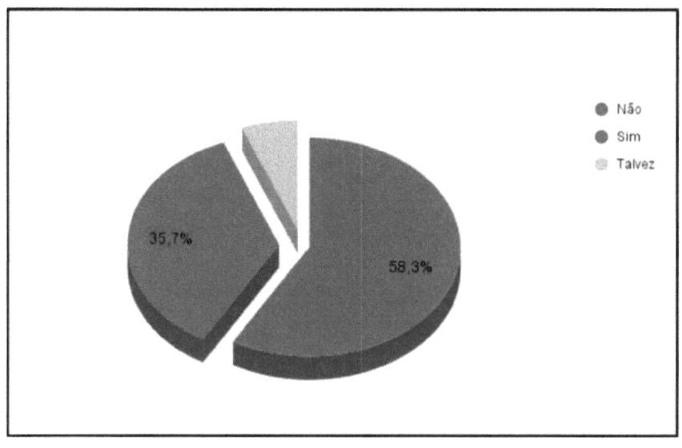

Source: Authorship Own (2023).

This practice is very current and according to the results, 58.3% of participants responded that they had not had their data cloned, however 35.7% responded that they had already fallen for this form of scam and curiously 6.0% were not convinced if they had already had their data. cloned.

Graphic 2.7 – Has anyone ever tried if to spend per you same (Blow of Whatsapp) it is per telephone, email or even physically in order to obtain advantages or privileged information?

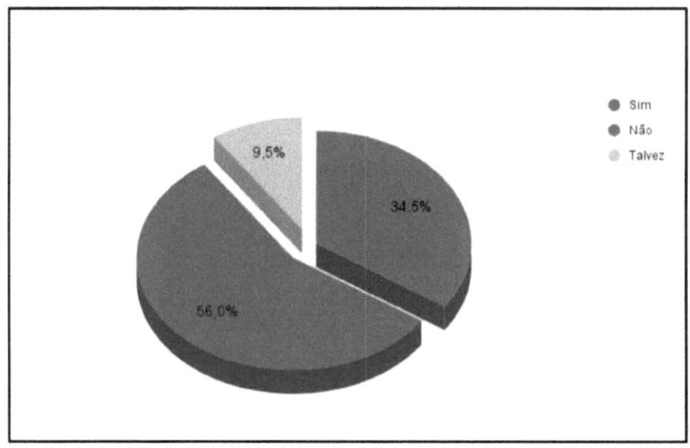

Source: Authorship Own (2023).

Another very current practice is the WhatsApp scam, in which attackers target known or even unknown people, but with the same intention in deceive The victim. In this context, according to you results 55.4% of the participants responded that they had not experienced such a situation, however 34.9% responded that they had already fallen for this form of scam.

Graphic 2.8 – Already typed information confidential in websites no insurance what caused damage psychological or financial?

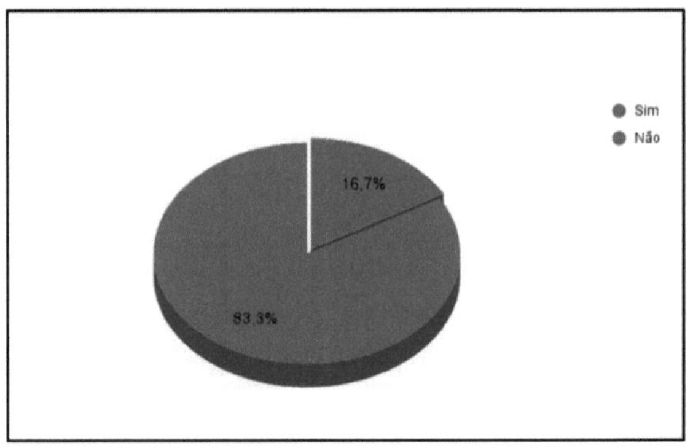

Source: Authorship Own (2023).

The majority, with 83.3%, of users Those surveyed responded that they had no losses with confidential information entered. On the other hand, a good number with 16.7% of the users unfortunately suffered some damage it is financial or until even psychological. A follow It is in fundamental importance The analysis of the results. hus, the research emphasized that a good part of the participants base their level of information on just basic knowledge about Social Engineering or have heard something in the media about this important topic, and the majority of respondents say they don't even know each other. In this context, participants have an adequate degree of knowledge in all frauds questioned, It is conscience in your importance stating to be one subject serious The be discussed. Furthermore, how much to the types in cheats known, you participants, reveal that they received the majority of the attacks asked. The results of this research brought to light some old problems, such as the high degree of false emails received, although few were financially affected. Furthermore, the old practice of prank calls involving prizes and kidnappings, where the vast majority claim to have received a false call despite the large part is aware of the risks, moreover, to extensive practice from sending SMS with links to fake draws, prizes, revealing the user's deception and curiosity and despite this, few were actually affected, to the theft of information, clone of documents in which the WhatsApp scam has a great impact and finally the loss of data or financial loss suffered by engineering Social in some strand, where 36% of the participants stated have been reached. It is well known that attacks that pique users' curiosity so that they click on news and images are surprisingly effective, in addition to hidden links where users must click on a link to continue on the page or to view such content. In short, it can be deduced from the research that the damage caused by social engineering is real and worrying, in which the user must always pay attention and be suspicious of any emails or news on social networks that appear suspicious, and always maintain their antivirus system updated. No there is doubts in what The attitude of participant It is The form more simple in prevent, primarily, a social engineering attack, from avoiding sharing personal information to always guaranteeing the veracity of all information researched. A follow one line of time in set with your Explanation will be presented in order to share maximum knowledge. Next, the main principles and reasons that permeate social engineering.

Figure 3: Line of Time Engineering Social

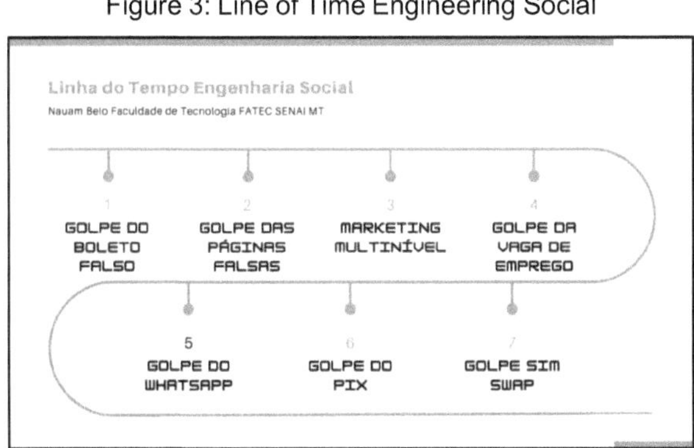

Source: Authorship Own, 2024

Figure 4: Explanation Line of Time Engineering Social

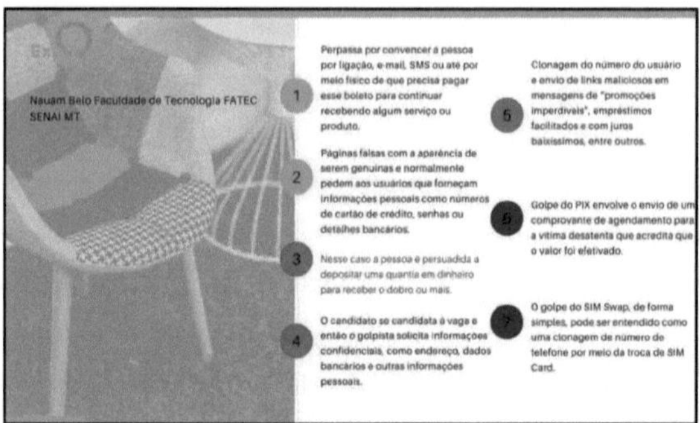

Source: Authorship Own, 2024

Figure 5: Seven Principles from the Engineering Social

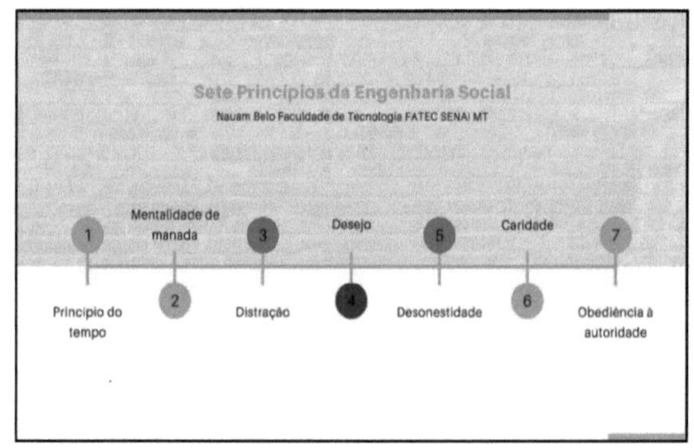

Source: Own Authorship, 2024

Figure 6: Main Reasons

Source: Authorship Own, 2024

There is no doubt that currently new scams appear in the blink of an eye, such line of time demonstrates The evolution of the attackers, in addition you reasons in which The population from the loopholes to the cybercriminals. Against of this It is necessary one action plan, in order to raise maximum awareness among the entire population. The action plan developed will be explained below.

According to a survey by PSafe, a company specializing in cybersecurity, between January and May 2022 alone, more than 3.4 million attempts in scams financial at the Brazil for the Internet – one average 22.5 thousand per day and around 930 per hour. In this context, Rodrigo Sgavioli Head of Allocation and Funds at XP Investimentos, explains: "The level of profitability, for example, of 3%, 4% or 5% per month without any risk is another indication. It is almost impossible for you to guarantee that, every month, you will have an investment five times more of what O return, without to take some risk". Therefore, case promises in high value are offered without any risk or investment, be wary it is a scam.

On October 16th, Science and Technology Day is celebrated in order to honor the great discoveries of men and also to encourage new scientific discoveries. Therefore, it is of fundamental importance not only to celebrate, but also to publicize it on radio stations and schools through awareness-raising booklets, in order to share as much knowledge as possible with those who have modest knowledge of technology or specific topics such as Social Engineering. With this, a booklet model as an action plan is then shared .

Figure 7: Disclosure Flat in action

Source: Authorship Own, 2024

Figure 8: Disclosure Flat in action Solution

Source: Authorship Own, 2024

5. CONSIDERATIONS FINALS

Therefore, through in all to the bibliographies studied for O development of this work, It was possible to conclude that the lack of necessary knowledge regarding Social Engineering specifically for the elderly generates negative impacts in such a digital world, so it is of fundamental importance to mitigate such risks. In this context, the objectives achieved in this study involve contextualization and awareness in such theme through in researches qualitative. Of that form, we can answer The Following question: As we can alert people elderly, what have low technological knowledge in relation to the negative impacts arising from vulnerability in the virtual environment?, through projects that aim at the continued training of people, as they can trigger skills and abilities for to guarantee one better security in all at the environment digital It is, thus, implementing good practices in using the Internet safely. Furthermore, based on the qualitative and quantitative research carried out, it can be stated that the elderly group is one of the favorite focuses of cybercriminals, therefore, teaching and raising awareness among the elderly to have good habits on digital networks is of paramount importance. A disclosure in measures effective It is O main goal, then how much The more the general population has this knowledge, the safer they will be to combat these cyber threats. In general, everyone who participated in this study demonstrated an interest in the topic of the work and in contributing with their information, providing them with greater and updated knowledge of the subject. about you main scratchs what The Internet provides for those ones what do not have any effective instruction, in addition it was possible to notice with the help of research, that together with the elderly, children are also the preferred targets of attackers, since they take advantage of such naivety on the part of the little ones and also often take some decision without the necessary support from those responsible. In short, the expected social effect will be a society capable of face the cyber crimes, through of knowledge acquired in studies such as this and future work that will be carried out, as it is recommended that longitudinal studies be carried out that prospectively assess the risks of Social Engineering in the digital era.

6. REFERENCES BIBLIOGRAPHICS

POSTGRADUATE CYBERSECURITY – CS20 Douglas Oliveira Fábio Zaguetto Juliano Santos
< https ://www.artig onal.com.br/hackeando-pessoas-atraves-da-engenharia-social/ >
Accessed on: 12 Jan. 2024.

Santos, Daniel Pitanga of the. "A engineering Social at the Brazil It is your risks." (2016).< http://riut.utfpr.edu.br/jspui/bitstream/1/19455/1/CT_GETIC_V_2015_05.pdf >
Accessed on: 20 Jan. 2024.

Behavioral analysis on social engineering attacks .Diss. Instituto Politecnico do Porto (Portugal), 2015. Gaspar, Jana Eça Hohlenwerger Muniz.
< https://w ww.proq uest.com/openview/c336a f 7f8bc824b3a7c0a7040a779c97/1?pq-origsite=gscholar&cbl=2026366&diss=y >
Access in: 15 Jan. 2024.

Social engineering in online social networks. Mato Grosso: State University of Mato Grosso, 2014. SANTOS, Yuri Rafael de Lima.
< https://w ww .academi a.ed u/39700176/A_ENGENHARIA_SOCIAL _
NAS_REDES_S OCAIS_ONLINE20190626_9659_zawg2a >
Access at: 05 Feb. 2024.

BATISTA, FL Methods and practices used in social engineering with the aim of preventing the theft of sensitive information. Brasília: Centro Universitário de Brasília, 2015.< https://repositorio.uniceub.br/jspui/bitstream/235/8155/1/51306378.pdf >
Accessed on: 06 Mar. 2024.

Bonini, Deise Mara Soares, et al. "Financial protection for the elderly in light of general data protection law." Research, Society and Development 10.12 (2021): e575101220973-e575101220973.
< https://rsdjournal.org/index.php/rsd/article/view/20973 >

Access in: 07 Mar. 2024.

Engineering Social: O what It is, types in attack, techniques It is as if protect < https://blogbr.clear.sale/engenharia-social-o-que-ee-como-se-protect#:~:text=Ou%20seja%2C%20a%20engenharia%20social,a%20partir%20da %20manipula%C3%A7%C3%A3o%20psicol%C3%B3gica .>

Access at: 17 April..2024.

CGI - COMMITTEE MANAGER FROM THE INTERNET AT THE BRAZIL< https://cgi.br/ >Access in: 20 April.. 2024.

Social-engineering-primer Available in:

< https://w ww.gov .br/abin/pt-br/acesso-a-informacao/acoes- e-

>programs/PNPC/good practices/booklet-social-engineering-guide-for-protection-of-sensitive-knowledgeAccess in: 20 April.. 2024.

CERT.BR. Internet Security Guide. Internet Steering Committee in Brazil, 2012. Available at: < https: //www .c gi.br/media/docs/publicacoes/1/cartilha- security-internet.pdf >
Access at: 21 April..2024.

CGI - COMMITTEE MANAGER FROM THE INTERNET AT THE BRAZIL

< https://cetic.br/pt/noticia/cresce-o-uso-de-internet-durante-a-pandemia-e-numero- of-users-in-brazil-reaches-152-millions-and what-the-skeptic-research-points to br/#:~:text=do%20Cetic.br-

,Grows%20%20use%20of%20Internet%20during%20a%20pandemic%20and%20n

%C3%B >Access at: 21 April.. 2024.

Cybersecurity in the new digital world: how to warn the elderly about the cyber risks resulting from phishing when using smartphones. Silva, Pedro Henrique da < https://repositorio.ufersa.edu.br/handle/prefix/8860 >
Access in: May 1st. 2024.

CETIC- center Regional in Studies for O Development from the Society of Information .
< https://cetic.br/pt/noticia/92-milhoes-de-brasileiros-acessam-a-internet- > just- pelo-telefone-celular-aponta-tic-domicilios-2022/
Access in: May 1st. 2024.

CNDL Number of elderly people accessing the internet grows from 68% to 97%, points out CNDL/SPC Brazil research. CNDL. 2021. < https://site.cndl.org.br/numero-de-idosos- who-access-the-internet-grows-from-68-to-

97-aponta-research-cndlspc-brasil/ > Accessed on: May 5th. 2024.

CONTEH, Nabie Y. ; SCHMICK, Paul J. Cybersecurity: risks, vulnerabilities and countermeasures to prevent social engineering attacks. EUA: ACCENTS, 2016. <https://scholar.google.com.br/scholar?hl=pt-BR&as_sdt=0%2C5&q=CONTEH%2C+Nabie+Y.+%3B+SCHMICK%2C+Paul+J.+Cybersecurity%3A+risks%2C+vulnerabilities+and+countermeasures+to+prevent+social

+engineering+attacks.+EUA%3A+ACCENTS%2C+2016.&btnG= > Accessed on: May 15th. 2024.

In Sa Barros, Solange Duarte Palm, and Paula Torales Milk. "A third age facing the challenges posed by technology: the need for learning for ethical and safe use." (2019).
< https://www.researchgate.net/profile/Solange-Barros-5/publication/335038225_A_terceira_idade_frente_aos_desafios_impostos_pela_tecnologia_a_necessidade_do_aprendizado_para_um_uso_etico_e_seguro/links/5d4c15e0a6fdcc370a85f255/The-third-age-face-with-the-challenges-imposed-by-technology-the-need-of-learning-for-ethical-and-safe-use.pdf >
Access in: May 15th. 2024.

"Engineering Social (or O ram what in the end it was one Wolf)." (2013). Country, Ricardo, Fernando Moreira, and João Varajão.
< https://repositorio.upt.pt/server/api/core/bitstreams/275c75cc-1546-41af-a9cd-1f4a52631f53/content >

Access in: May 15th. 2024.

Brazil registered more in 234 millions in hits furniture in 2020
GOV.BR.< https:/ /www . gov.br/pt-br/noticias/transito-e-transportes/2021/05/brasil-registered-more-than-234-million-mobile-accesses-in-2020 >

Access in: May 20th. 2024.

GUEDES, M. S.; CHICK, A. A. N.; SON, A. A. B.; BIRTH, P. H.; KINGS, D. L.;

CEDRAN, PC; COSTA, AP Virtual crimes and scams: challenges faced by the elderly in the technological age. LATINOAMERICAN ECONOMY OBSERVATORY, [S.l.] , v. 21,n.9,P.14026–14040 2023.IT HURTS:
10.55905/oelv21n9- 190. Available at:

< https://ojs.observatoriolatinoamericano.com/ojs/index.php/olel/article/view/129 >3. Accessed on: 20 Mar. 2024.

Is aware: blow in phishing involving offers in job fraudulent

< HTTPS :// WWW . IBM . COM / BR - PT / CAREERS / PHISHING - SCAMS > Access in: 10 sea. 2024. What is phishing?

< HTTPS :// WWW . IBM . COM / BR - PT / TOPICS / PHISHING > Access in: 10 Mar. 2024.

"Phishing during the Covid-19 pandemic." (2021). RIBEIRO, Angelo Henrique Lorenzi, and Renan dos Santos CONCORDIA.
< https://ric.cps.sp.gov.br/handle/123456789/12792 > Accessed on: May 23. 2024.

PEREIRA, Cleber Guedes. Phishing: Concepts and preventive actions applied to the company. Brasília, 2012. 56 p Course Completion Work (Computer Networks) - center University in Brasilia, Brasilia, 2012. Available in:
< https://blog.grupogen.com.br/juridico/postagens/dicas/ataques-e-crimes- cybernetics/ >Access in: 23 May. 2024.

HIGHLAND. LAW NO. 10,741, OF 1st IN OCTOBER OF 2003. PLANALTO: Presidency from the Republic Home Civil Deputy Chief for Matters Legal. 2003. Available in:

< http://www.planalto.gov.br/ccivil_03/LEIS/2003/L10.741compilado.htm > Accessed on: 23 May. 2024.
HIGHLAND. LAW No. 12,737, IN 30 IN NOVEMBER IN 2012. HIGHLAND: Law Carolina Dieckmann. Available in:

< https://w ww .planalto. gov .br/cciv il _03/_ato2011-2014/2012/lei/l12737.ht m > Accessed on: 25 May. 2024.

HIGHLAND. LAW No. 13,709, IN 14 IN AUGUST IN 2018. HIGHLAND: Law General from the protection in data. Available in:

< https://w ww .planalto. gov .br/cciv il _03/_ato2015-2018/2018/lei/l13709.ht m > Accessed on: 25 May. 2024.

Pity, Brian Henrique, Anderson Santos from the Silva, and Maicon of the Saints. "PHISHING." MANAGEMENT AND EDUCATION TECHNOLOGY SEMINAR 2.2 (2020).

< https://repositorio.pucgoias.edu.br/jspui/bitstream/123456789/6206/1/Artigo%20-%20Sabrina%20Moreno.pdf > Accessed on: May 26th. 2024.

Pereira, Cleber Guedes."Phishing: Concepts It is actions preventive applied to the company." (2012).
< https://repositorio.uniceub.br/jspui/handle/235/8136 > Accessed on: 26 May. 2024.

"Phishing." Magazine from the Faculty in Right from the University in Lisbon 54 (2013): 87-102. Gerald, A-N-A Vaca.

< https://repositorio.ul.pt/handle/10451/59340 > Accessed on: 26 May. 2024.

"Phishing in the information age: relevance of personal data protection." (2022). Marichal, Pedro Luiz de.
< https://lume.ufrgs.br/handle/10183/258868 > Accessed on: 26 May. 2024.

"PHISHING AND ENGINEERING SOCIAL: IN BETWEEN A CRIMINALIZATION AND AUSE OF PROTECTIVE SOCIAL MEDIA." Meritum : Revista de Direito da Universidade FUMEC 15.1 (2020). de Oliveira Fornasier, Mateus, Norberto Milton Paiva Knebel, and Fernanda Viero da Silva.
< https://openurl.ebsco.com/EPDB%3Agcd%3A7%3A25338615/detailv2?sid=ebsco%3Aplink%3Ascholar&id=ebsco%3Agcd%3A145674477&crl=c >
Access in: May 27th. 2024.

"Protection in data It is O study from the GDPR." (2021). burkart, Daniele Vincenzi Villares.< https://repositorio.unesp.br/items/d3f31333-1765-4c9b-998e-036640aee715 > Accessed on: 27 May. 2024.

Ransomware: defending yourself from digital extortion. Novatec Editora, 2019. Liska, Allan, and Timothy Gallo.
<https://books.google.com.br/books?hl=pt-BR&lr=&id=gf6ZDwAAQBAJ&oi=fnd&pg=PT4&dq=Ransomware:+defendendo-se+da+extors%C3%A3o+digital.+Novatec+Editora,+2019.+Liska,+Allan,+and+Timothy+Gallo.+&ots=SGZiEKgg4J&sig=b0-ySQT2_ukQby0bl6dnOv1Ji00#v=onepage&q=Ransomware%3A%20defendendo-se%20da%20extors%C3%A3o%20digital.%20Novatec%20Editora%2C%202019.%20Liska%2C%20Allan%2C%20and%20Timothy%20Gallo.&f=false>
Acesso em: 10 mar. 2024

"Ransomware It is cybersecurity: The information threatened per attacks The data." Revista Thesis Juris 9.1 (2020): 208-236.

in Oliveira Fornasier, Mateus, Tiago Protti Spinato, and Fernanda Lencina Ribeiro.

< https://periodicos.uninove.br/thesisjuris/article/view/16739 >

Access in: 10 Mar. 2024

from the Silva, A. v. O., Lime, Angela T. P., from the Silva, L. W., It is Wedge, L. D. L., & of Carmo,

A. S. (2023). PROTECTION AND SECURITY FOR O ELDERLY AT THE QUITE INFORMATION TECHNOLOGY. Contemporary Magazine , 3 (12), 29922–29938.
< https://doi.org/10.56083/RCV3N12-260 > Accessed on: 10 March. 2024

SCARPIONI, Agesandro, et al. "Development of a virtual environment for training elderly people to avoid Internet scams." ESPACIOS Magazine | Vol. 37 (No. 09) Year 2016 (2016).
< https://w ww.rev istaespacios.com/ a1 6v37n09/16370913.html > Accessed on: May 1st. 2024
Take cover against phishing
< **HTTPS :// SUPPORT. MICROSOFT . COM / PT - BR / WINDOWS / PROTEJA - SE - CONTRA - PHISHING - 0 C 7 EA 947- BA 98-3 BD 9-7184-430 E 1 F 860 A 44 >**
Access in: 11 May. 2024
Aging today: chronological, biological, psychological and social aspects
< https://w ww .scielo.br/j/estpsi/a /LTdt hHbL v ZPLZk8MtMNmZyb/abstract/?lang=pt >Access in: May 28th. 2024
"Technique phishing: simple, but effective." (2017). Would make, Thiago Stefanini.

< https://ric.cps.sp.gov.br/handle/123456789/773 >

Access in: May 28th. 2024

WOJAHN, A. S. .; MICHAEL, c. gives Q. .; VEIGA, d. JS gives; LENZ, R. .; SILVA, S. g.da .; ROSSETTO, T. P. .; SANTOS, M. L. dos . The social vulnerability of the elderly against scams in the digital scope. Research, Society and Development, [S. l.], v. 11,n. 11, P. e452111133652, 2022. IT HURTS: 10.33448/rsd-v11i11.33652. Available in:https://rsdjournal.org/index.php/rsd/article/view/33652. Accessed on: 20 Mar. 2024.

APPENDIX A

SEARCH IN ASSESSMENT ABOUT O LEVEL IN KNOWLEDGE AND INTERNET SAFETY AMONG THE POPULATION

- That search he was developed as part of work in conclusion of analysis course and Development in Systems from the Faculty in Technology of State of Mato Grosso (FATEC - SENAI) and aims to identify the level of knowledge from the population about O use It is The security from the Internet, to the which are exposed on a daily basis.

- Objective: Analyze as many responses as possible in order to obtain essential information.

- Answers to these questions should be given honestly, without letting other people's opinions influence your answers, just based on your current knowledge and behavior. The questionnaire will take around 10 minutes to complete.

- Such quiz It is Divided in two parts: questions general It is specific.

- To ensure respondents' privacy, we will not ask for any information that could identify them, such as name or contact information.

Pesquisa de Avaliação sobre o nível de Conhecimento e a Segurança da Internet entre a população.

- Essa pesquisa foi desenvolvida como parte do trabalho de conclusão do curso de Análise e Desenvolvimento de Sistemas da Faculdade de Tecnologia do Estado de Mato Grosso (FATEC - SENAI) e tem por objetivo identificar o grau de conhecimento da população sobre o uso e a segurança da Internet, aos quais estão expostos no dia a dia.
- Objetivo: Analisar o maior número de respostas, a fim de obter informações essenciais.
- As respostas a essas perguntas devem ser dadas honestamente, sem deixar que as opiniões de outras pessoas influenciem suas respostas, apenas com base em seu conhecimento e comportamento atual. O questionário levará em torno de 10 minutos para ser respondida.
- Tal questionário é dividido em duas partes perguntas gerais e específicas.
- Para garantir a privacidade dos respondentes, não solicitaremos nenhuma informação que possa identificá-los, como nome ou informações de contato.

Contextualizando tal pesquisa, você já recebeu links desconhecidos com promessas de altas recompensas ou bancos pedindo informações confidenciais? Tais técnicas são algumas entre milhares que os denominados "atacantes" utilizam para conseguir o máximo proveito de pessoas com conhecimentos tecnológicos modestos. Neste Contexto, entra um dos assuntos mais populares com avanço da internet a Engenharia Social, uma vez que perpassa por enganar as pessoas utilizando o elo mais fraco dos mecanismos de segurança o próprio ser humano.

Idade *

12 a 18 anos

19 a 25 anos

26 a 36 anos

Acima de 37 anos

Gênero *

Masculino

Feminino

Escolaridade *

Ensino Fundamental

Ensino Superior

Ensino Médio

Nenhum

Questões da Pesquisa: Perguntas Gerais!!!
Nesta seção você deve assinalar somente uma das respostas para cada pergunta.

Você utiliza a Internet? *

Diversas vezes ao dia

Pelo menos uma vez ao dia

Pelo menos uma vez por semana

Não uso internet

O que você costuma usar para acessar a internet? *

Computador de mesa

Notebook

Celular

Tablet

Outros...

Qual desses aplicativos você costuma acessar para se comunicar com alguém da família? *

SMS

Fazer ligação

WhatsApp

Facebook

Instagram

Quando você tem alguma dificuldade de entender mensagens ou ligações, você costuma pedir ajuda a alguém? *

- [ ] Sim

- [ ] Não

Caso alguém ligue ou envie uma mensagem informando que é um funcionário de algum banco e peça seu nome completo, número do seu CPF, número do telefone, número da conta e a senha, você passaria? *

- [ ] Sim

- [ ] Pediria ajuda de alguém da minha confiança

- [ ] Não

Você acha que o acesso da Internet oferece algum perigo? *

- [ ] Sim

- [ ] Não

Você acha que pesquisas como essa ajudam na conscientização de boas práticas na segurança da Internet, a fim de identificar possíveis golpes? *

- [ ] Sim

- [ ] Não

- [ ] Talvez

Questões da Pesquisa: Perguntas Específicas!!!

Nesta seção você deve assinalar somente uma das respostas para cada pergunta.

Qual é o seu nível de conhecimento sobre o tema 'Engenharia Social'? *

- [ ] Total conhecimento e domínio do assunto

- [ ] Possuo algum conhecimento

- [ ] Já ouvi falar pela mídia

- [ ] Nenhum

Na sua opinião quais vítimas são mais propícias em ser enganadas? *

- [ ] Idosos

- [ ] Adolescentes

- [ ] Adultos

- [ ] Crianças

- [ ] Outros...

Já recebeu um e-mail falso? *

- [ ] Sim

- [ ] Não

Caso tenha recebido um e-mail falso: Chegou a clicar no link ou anexo nele contido, por curiosidade? *

☐ Sim

☐ Não

Já clicou em um link de forma enganosa enquanto navegava em algum site? *

☐ Sim

☐ Não

Já digitou informações confidenciais em sites não seguros que causaram danos de alguma forma? *

☐ Sim

☐ Não

Alguma vez você já recebeu um telefonema com uma tentativa de fraude sobre um sequestro, prêmio ou banco...? *

☐ Sim

☐ Não

Caso tenha recebido a ligação de um fraudador, chegou a realizar o que foi pedido? *

☐ Sim

☐ Não

Já recebeu SMS falso de um sorteio que tenha ganho um prêmio? *

☐ Sim

☐ Não

☐ Talvez

Em algum momento já teve seus dados clonados seja cartão de crédito ou dados pessoais? *

☐ Sim

☐ Não

☐ Talvez

Alguém já tentou se passar por você mesmo (Golpe do Whatsapp) seja por telefone, e-mail ou até mesmo fisicamente a fim de conseguir vantagens ou informações privilegiadas? *

☐ Sim

☐ Não

☐ Talvez